STARTING OUT

BULLET TRAINS

MEG GREVE

CREATIVE EDUCATION • CREATIVE PAPERBACKS

CONT

ENTS

I See a Bullet Train 4

All Aboard 6

A Quiet Ride 8

Floating Trains 10

Make a Noise 12

Bullet Train Words 14

Reading Corner 15

Index 16

I SEE A BULLET TRAIN.

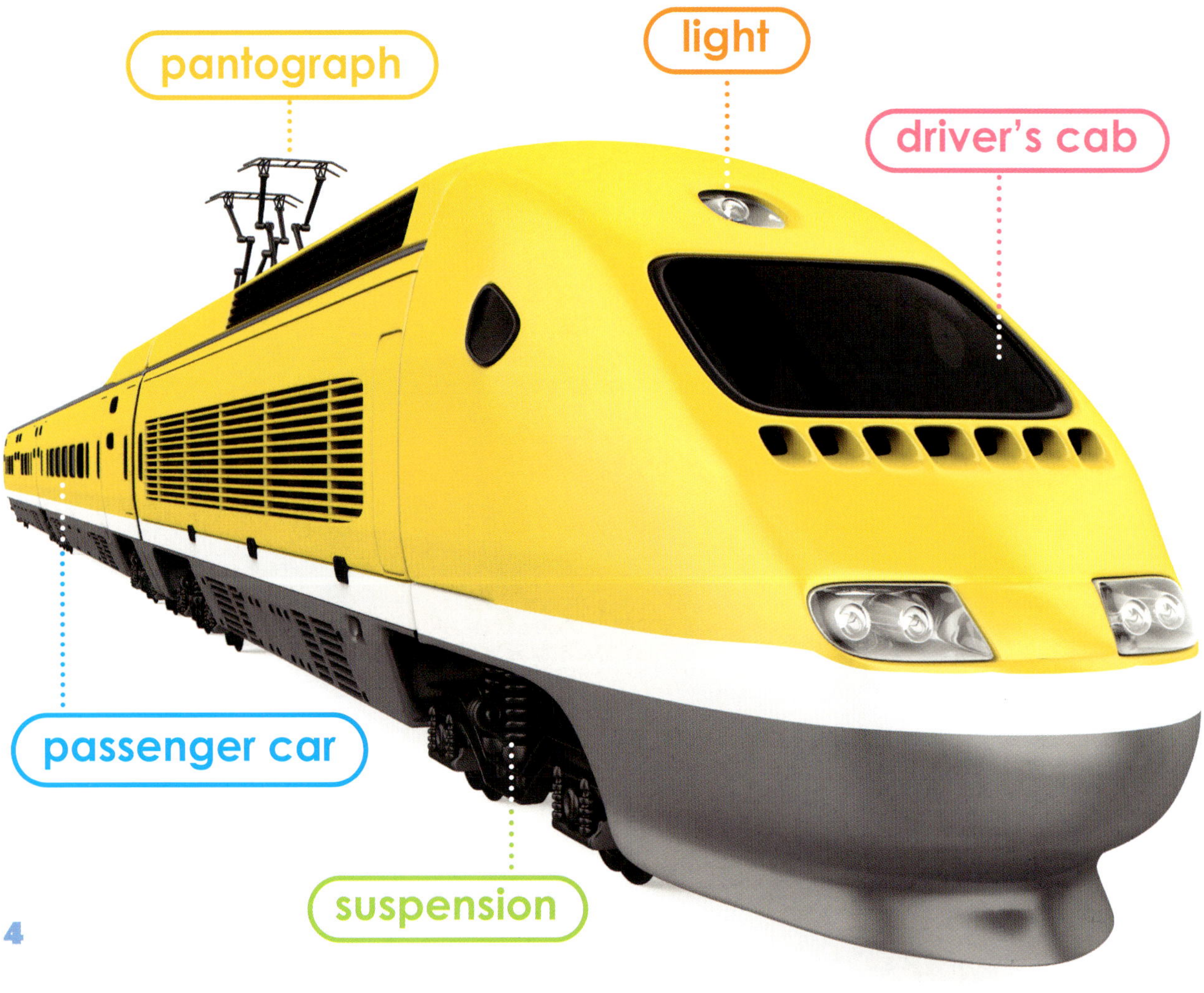

It looks blurry as it whizzes past.

This is one of the fastest trains in the world.

16

Most trains carry things. Bullet trains carry passengers.

Airplanes go fast too. But trains can fit more people.

Most bullet trains have electric lines over the track. A bullet train runs on electricity.

The ride is smooth and quiet.

Electricity goes between the magnets.

Some bullet trains have big magnets. Magnets are in the track too.

The train floats above the track.

MAKE A NOISE

OOSSH!

Can you make the sounds of a bullet train?

Listen to these sounds:

https://www.youtube.com/watch?v=qDkbl5xzlFA

Now it is your turn!

BULLET TRAIN WORDS

electricity: a form of energy; energy that is often carried through wires

magnets: pieces of metal that pull in or push away other magnets

passengers: people who ride on a train or other vehicle

READING CORNER

Holzweiss, Kristina A. *My First Book of Trains: All About Locomotive and Railcars for Kids.* New York: Rockridge Press, 2022.

Murray, Julie. *Shinkansen Bullet Train*. Edina, Minn.: Dash!, 2022.

Wing, Scarlett. *Trains Then and Now*. Rolling Meadows, Ill.: Cottage Door Press, 2019.

INDEX

airplanes 7

carry 7

electricity 8, 10

fast(est) 5, 7

floats 11

magnets 10, 11

people 7

ride 9

track 8, 11

PUBLISHED BY CREATIVE EDUCATION AND CREATIVE PAPERBACKS
P.O. Box 227, Mankato, Minnesota 56002
Creative Education and Creative Paperbacks are imprints of The Creative Company
www.thecreativecompany.us

LIBRARY OF CONGRESS CATALOGING-IN-PUBLICATION DATA
Names: Greve, Meg, author.
Title: Bullet trains / by Meg Greve.
Description: Mankato, Minnesota : Creative Education and Creative Paperbacks, [2025] | Series: Starting out | Includes bibliographical references and index. | Audience: Ages 4-7 | Audience: Grades K-1 | Summary: "Bullet trains will introduce budding book learners to a noisy, colorful world with this new Starting Out title. Colorful photos, labeled diagrams, 'Make a Noise' section, glossary, and more ignite a passion for learning"-- Provided by publisher.
Identifiers: LCCN 2023059418 (print) | LCCN 2023059419 (ebook) | ISBN 9798889891680 (library binding) | ISBN 9781682775530 (paperback) | ISBN 9798889891802 (ebook)
Subjects: LCSH: High speed trains--Juvenile literature. | CYAC: High speed trains. | LCGFT: Instructional and educational works.
Classification: LCC TF1455 .G74 2025 (print) | LCC TF1455 (ebook) | DDC 625.2--dc23/eng/20240131
LC record available at https://lccn.loc.gov/2023059418
LC ebook record available at https://lccn.loc.gov/2023059419

DESIGN AND PRODUCTION
Design by Rhea Magaro and Jennifer Bowers
Production by Beeline Media and Design
Art direction by Tom Morgan
Printed in the United States of America

PHOTOGRAPHS by Dreamstime (© Meinzahn), Getty Images (3alexd, baona, BanksPhotos), Shutterstock (Nerthuz, freestore 839, Nerthuz, yuyangc, phive, Sean Pavone, Artur Rodin, Oleksiy Mark, Bohbeh, aappp)